ACCIDENTAL SCIENTIFIC DISCOVERIES THAT CHANGED THE WORLD

OOPS! IT'S DYNAMITE!

BY MELISSA RAÉ SHOFNER

Gareth Stevens PUBLISHING

Please visit our website, www.garethstevens.com. For a free color catalog of all our high-quality books, call toll free 1-800-542-2595 or fax 1-877-542-2596.

Library of Congress Cataloging-in-Publication Data

Names: Shofner, Melissa Raé, author.
Title: Oops! It's dynamite! / Melissa Raé Shofner.
Other titles: Oops! It is dynamite | It's dynamite | It is dynamite
Description: New York : Gareth Stevens Publishing, [2020] | Series: Accidental scientific discoveries that changed the world | Includes bibliographical references and index.
Identifiers: LCCN 2018044860| ISBN 9781538239865 (pbk.) | ISBN 9781538239889 (library bound) | ISBN 9781538239872 (6 pack)
Subjects: LCSH: Nobel, Alfred Bernhard, 1833-1896–Juvenile literature. | Dynamite–Juvenile literature. | Explosives–Juvenile literature. | Discoveries in science–Juvenile literature. | Chemical engineers–Sweden–Biography–Juvenile literature.
Classification: LCC TP268.5.N7 S54 2020 | DDC 662/.27–dc23
LC record available at https://lccn.loc.gov/2018044860

First Edition

Published in 2020 by
Gareth Stevens Publishing
111 East 14th Street, Suite 349
New York, NY 10003

Designer: Katelyn E. Reynolds
Editor: Monika Davies

Photo credits: Cover, p. 1 Jeffrey Coolidge/The Image Bank/Getty Images; cover, pp. 1–32 (burst) jirawat phueksriphan/Shutterstock.com; cover, pp. 1–32 (burst lines) KID_A/Shutterstock.com; p. 5 (main) George Rinhart/Corbis via Getty Images; p. 5 (inset) canadastock/Shutterstock.com; p. 7 Adam Burn/ Getty Images; p. 8 Serg64/Shutterstock.com; p. 9 English Heritage/Heritage Images/Getty Images; p. 11 (main) SSPL/Getty Images; p. 11 (inset) Mark Oniffrey/Intropin/Wikipedia.org; p. 13 apiguide/ Shutterstock.com; p. 15 Photo 12/UIG via Getty Images; p. 17 Jason Rothe/Shutterstock.com; p. 19 (main) achristopher73/Shutterstock.com; p. 19 (inset) ggw/Shutterstock.com; p. 21 Pbroks13/Wikipedia.org; p. 23 Fine Art Images/Heritage Images/Getty Images; p. 24 CSLD/Shutterstock.com; p. 25 Andrey N Bannov/Shutterstock.com; p. 27(top left) PHILIPPE DESMAZES/AFP/Getty Images; p. 27 (top right) Sirocco/Shutterstock.com; p. 27 (bottom) Roger de Montfort/Shutterstock.com; p. 29 Evans/Hulton Archive/Getty Images.

Printed in the United States of America

CPSIA compliance information: Batch #CS19GS: For further information contact Gareth Stevens, New York, New York at 1-800-542-2595.

CONTENTS

Words in the glossary appear in bold type the first time they are used in the text.

AMAZING ACCIDENTS

People use science to help them better understand and explain the world around them. Scientists have learned a lot by studying the natural world and testing their theories, or ideas, through carefully planned experiments. But sometimes, amazing scientific discoveries are made by accident!

Some of the most useful and important scientific discoveries in history have been made when scientists were looking for something else—or, in some cases, not looking for anything at all! One such accident over 150 years ago led to the invention of dynamite. Believe it or not, this explosive discovery actually made the science of blowing stuff up safer.

EXPLOSIVES EXPLAINED

EXPLOSIVES CAN BE **MECHANICAL**, NUCLEAR, OR CHEMICAL IN NATURE. CHEMICAL EXPLOSIVES, SUCH AS DYNAMITE, ARE MOST COMMON. THEY'RE MADE OF MIXTURES OR COMPOUNDS THAT BREAK DOWN OR REARRANGE VERY QUICKLY WHEN HEATED OR SHOCKED. WHEN THIS HAPPENS, LARGE AMOUNTS OF GAS AND HEAT ARE RELEASED, OR LET GO, IN AN EXPLOSION.

The Hoover Dam is located on the Nevada-Arizona border. While the dam was being built in the early 1930s, water from the Colorado River needed to be rerouted through four 56-foot (17 m) tunnels. Dynamite made the job much easier for **engineers**. Imagine digging those tunnels by hand!

AN EXPLOSIVE HISTORY

To understand why the invention of dynamite was so important, we should first look at the history of explosives. The earliest-known chemical explosive, black powder, was first used in China around the 9th century.

Black powder, also called gunpowder, needs heat or flame to ignite, or catch fire. A single spark is all it takes! Fuses allowed people to ignite the powder from far away, but it was hard to tell exactly when it would catch fire. Many people were killed. The safety fuse was invented in 1831. This new type of fuse burned at a steady pace, making it much safer to ignite black powder.

The Right Mix

BLACK POWDER IS A VERY SMOKY EXPLOSIVE. IT'S A MIXTURE OF THREE INGREDIENTS: SALTPETER, CHARCOAL, AND SULFUR. THE AMOUNT OF EACH INGREDIENT NEEDS TO BE JUST RIGHT FOR THE BLACK POWDER TO WORK. CHANGING THE AMOUNTS AFFECTS HOW THE POWDER BURNS AND THE AMOUNT OF SMOKE IT CREATES.

Black powder was first used in fireworks and signals. It was later used in weapons such as guns and bombs, and for mining and construction. Today, black powder is mostly used in fireworks.

In 1846, German **chemist** Christian Schönbein spilled some nitric and sulfuric acid in his kitchen and wiped up the mess with a cotton apron. After the apron dried, it suddenly burst into flames. The explosion didn't create much heat or smoke.

Schönbein had discovered nitrocellulose, also called guncotton, a chemical explosive nearly six times as powerful as black powder. Scientists soon learned how to safely make guncotton and began using it in weapons. Within a year, however, they realized guncotton would randomly explode if it wasn't kept cool, dry, and clean. This made it less popular and people stopped using it.

OLD FILM REEL

A Surprising New Use

SCIENTISTS EVENTUALLY FIGURED OUT THAT NITROCELLULOSE COULD BE PUT TO OTHER USES IF IT WAS PLACED INSIDE ANOTHER SUBSTANCE, SUCH AS WAX. IN THIS WAY, IT WAS USED TO MAKE FILM FOR CAMERAS. THIS FILM WAS FIRST USED FOR PICTURES AND LATER MOVIES AND X-RAYS, BUT IT WAS STILL HIGHLY **FLAMMABLE**.

Movie theaters, such as this one, posted "No Smoking" signs in areas where the highly flammable film was used, but fires still happened—and they couldn't be put out with water! Companies stopped making film with nitrocellulose in the 1950s.

NITROGLYCERIN

Around the time that nitrocellulose was discovered in Germany, an Italian chemist named Ascanio Sobrero created a new chemical explosive called nitroglycerin. In nitroglycerin, nitric and sulfuric acid are mixed with glycerol to create an oily, colorless liquid. It's very unstable, which means it has a chemical structure or physical state that changes easily.

Remember earlier when we learned that a changing state is what causes chemical explosives to release huge amounts of energy in the form of an explosion? Nitroglycerin is highly **volatile**—even the slightest bump can cause it to explode! This makes it very unsafe to handle and transport, or move.

Missiles and Medicine

MANY CHEMICAL EXPLOSIVES CAN BE USED AS PROPELLANTS. PROPELLANTS ARE OBJECTS OR SUBSTANCES THAT CREATE ENERGY THAT PUSHES ANOTHER OBJECT FORWARD. NITROGLYCERIN AND NITROCELLULOSE ARE SOMETIMES COMBINED TO CREATE PROPELLANTS FOR ROCKETS AND WEAPONS. AMAZINGLY, THE LIQUID NITROGLYCERIN USED IN ROCKET FUEL IS ALSO USED AS A SOLID IN TINY AMOUNTS TO TREAT HEART CONDITIONS!

Ascanio Sobrero told others about the dangers of nitroglycerin, but his warnings didn't stop it from being used as an explosive.

NITROGLYCERIN TABLETS

THE SCIENCE OF EXPLOSIVES

Chemical explosives are divided into two groups: high explosives and low explosives. Low explosives, such as black powder, are sometimes called deflagrating explosives. This is because they deflagrate, which means they burn quickly but **subsonically** and create low pressures. In most cases, controlled explosions are examples of deflagration. Fire is also an example.

High explosives, such as nitrocellulose and nitroglycerin, are sometimes called detonating explosives. When something detonates, it explodes or causes something to explode. High explosives react **supersonically**, which makes them burn very fast and create high pressures. These explosions are sudden and extreme and may cause great damage.

Molecules Matter

LOW EXPLOSIVES CONTAIN **MOLECULES** OF A FUEL SOURCE AND AN **OXIDIZER.** THEY RELEASE THEIR ENERGY AS THE FUEL AND OXIDIZER MOLECULES COMBINE DURING **COMBUSTION.** IN HIGH EXPLOSIVES, THE FUEL AND OXIDIZER ARE ALREADY COMBINED INSIDE MOLECULES. THESE MOLECULES CONTAIN GREAT ENERGY THAT'S RELEASED ALL AT ONCE IN AN EXPLOSION IF THEY'RE BUMPED OR SHAKEN.

Low explosives can also cause explosions, but to do so they need to be contained. This is how fireworks work. Black powder is packed into a tube, and once it's all done burning, the gases that have built up burst out with a bang!

ALFRED NOBEL

Alfred Nobel was a Swedish chemist who understood the science behind chemical explosives. His father was an engineer who operated an explosives factory in Russia. Nobel studied chemistry in France and worked in the United States for several years before going to work for his father.

After his father's factory closed, Nobel continued to study explosives. Black powder was popular at the time, and nitroglycerin had recently been discovered. In 1862, Nobel built his own small factory and began to manufacture nitroglycerin. He knew of the dangers of handling such an unstable substance and began looking for ways to make it safer.

A Dangerous Business

NOBEL STARTED STUDYING NITROGLYCERIN WITH HIS FATHER AROUND 1859. THE FACTORY NOBEL BUILT IN 1862 EXPLODED IN 1864, KILLING SEVERAL PEOPLE, INCLUDING HIS YOUNGEST BROTHER. FOR A WHILE, NOBEL CONTINUED TO STUDY NITROGLYCERIN ON A BOAT IN THE MIDDLE OF A LAKE. HE LATER BUILT SEVERAL OTHER FACTORIES AROUND EUROPE.

Alfred Nobel was greatly affected by his brother's death and continued to look for ways to make nitroglycerin safer to handle.

Nobel invented many new devices in his lifetime. By the time he died in 1896, he owned 355 **patents**! One was for a small wooden device containing black powder that was placed inside a metal container of nitroglycerin. When the black powder was ignited, it would in turn detonate the nitroglycerin, providing a safer way to work with the volatile substance.

Nobel improved upon this device and patented the blasting cap in 1865. This invention marked the beginning of the modern use of high explosives. Today, many people in the explosives **industry** still consider the blasting cap to be the greatest invention since black powder.

Combining Inventions

TO SAFELY IGNITE NOBEL'S BLASTING CAP, THE DEVICE WAS COMBINED WITH THE SAFETY FUSE WILLIAM BICKFORD HAD INVENTED IN 1831. BICKFORD'S SAFETY FUSE WAS MADE FROM BLACK POWDER WRAPPED TIGHTLY INSIDE A NATURAL FIBER CALLED JUTE. OTHER FIBERS WERE SOMETIMES USED. TODAY'S SAFETY FUSES ARE VERY SIMILAR, THOUGH THEY'RE NOW WATERPROOFED AND SOMETIMES WRAPPED IN PLASTIC.

Nobel's blasting cap was used for over 50 years with few changes. This is what a blasting cap looks like today.

NOBEL'S ACCIDENTAL DISCOVERY

In 1867, Nobel made an incredible—and accidental—discovery. In one version of the story, he was transporting nitroglycerin, which was very dangerous since the slightest bump could cause an explosion. Disaster nearly occurred, too. Nobel noticed one of the containers had broken and nitroglycerin was leaking out!

The containers of nitroglycerin had been packed in kieselguhr, a type of rock mixture. To Nobel's surprise, the kieselguhr had absorbed, or taken in, the nitroglycerin and stopped it from exploding. Back in his lab, Nobel began to study the use of kieselguhr as a stabilizer for nitroglycerin. The result was his latest explosive invention—dynamite!

ANOTHER TELLING

SOME SAY NOBEL MADE HIS DISCOVERY AFTER DROPPING A CONTAINER OF NITROGLYCERIN IN HIS LAB. THE EXPLOSIVE LIQUID WAS ABSORBED BY SOMETHING ON THE FLOOR AND DIDN'T DETONATE. NOBEL EXPERIMENTED WITH DIFFERENT SUBSTANCES, LOOKING FOR ONE THAT WOULD STABILIZE NITROGLYCERIN WITHOUT MAKING IT LESS POWERFUL. HE DECIDED KIESELGUHR WAS THE BEST CHOICE.

Kieselguhr, also called diatomaceous earth, is a claylike substance. It contains the fossilized remains of diatoms, which are tiny organisms that live in the water.

DIATOMS

BOOM, DYNAMITE!

Kieselguhr seemed to be the answer to Nobel's questions about how to make nitroglycerin safer to handle and transport. He experimented further and found that adding 75 percent kieselguhr to 25 percent nitroglycerin was the perfect mix. The nitroglycerin was absorbed by the **silica** in the kieselguhr and formed a paste. The silica stopped the explosive from breaking down and detonating.

Nobel shaped his new, stable paste into short sticks, which he patented as dynamite in 1867. He improved his blasting caps and added them to the dynamite sticks so they could be ignited with a fuse.

SETTING UP SHOP

THE INVENTION OF DYNAMITE CREATED A NEW EXPLOSIVES INDUSTRY. OVER THE NEXT 10 YEARS, NOBEL, WITH THE HELP OF OTHERS, OPENED 16 NEW FACTORIES IN 14 COUNTRIES. IN 1886, SEVERAL OF THESE COMPANIES JOINED TOGETHER TO CREATE THE NOBEL DYNAMITE TRUST COMPANY. EVEN THOUGH NOBEL PATENTED HIS EXPLOSIVE INVENTIONS, OUTSIDE COMPANIES STILL STOLE HIS IDEAS.

Nobel chose the word "dynamite" based off the Greek word *dynamis*, which means "power." He later found that kieselguhr actually did take away some of nitroglycerin's explosive power, and he switched to using other absorbent substances, such as **wood pulp**, instead.

DYNAMITE

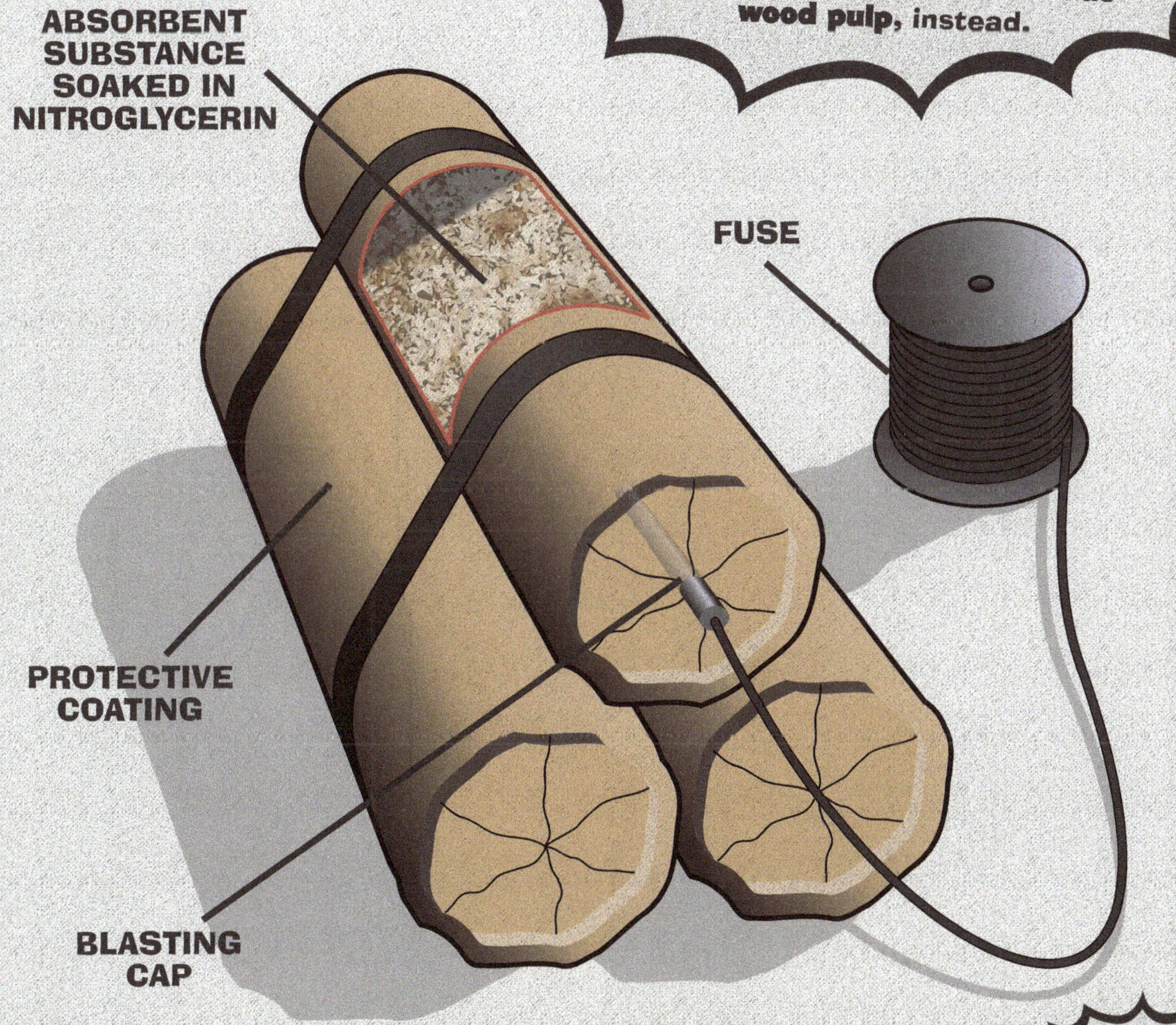

The invention of dynamite changed the world. Also called Nobel's Safety Powder, it provided miners, construction workers, and engineers with a safer way to do their jobs. They no longer had to worry about black powder igniting too soon or too late or surrounding them in smoke. They also didn't need to worry about handling and transporting dangerous nitroglycerin.

Dynamite became widely used in both construction and demolition, or the knocking down of buildings or other structures. It was used to blast away earth so tunnels, roads, canals, and dams could be built. Rocks that were also blasted out could then be reused in construction projects.

A Bigger Boom

IN 1875, NOBEL ACCIDENTALLY CREATED A NEW, MORE POWERFUL FORM OF DYNAMITE BY COMBINING NITROGLYCERIN WITH NITROCELLULOSE. THIS MIXTURE FORMED A PLASTIC-LIKE SUBSTANCE THAT WASN'T EASILY AFFECTED BY WATER. HE PATENTED THIS NEW EXPLOSIVE, WHICH HE CALLED BLASTING GELATIN, IN 1876. HE LATER ADDED OTHER INGREDIENTS TO THE MIX AND CALLED THE NEW MIXTURES "GELATIN DYNAMITES."

Nobel invested his money in other areas of the explosives industry. This means he put his money into other companies in order to make a profit later. These investments along with his own dynamite companies made Nobel rich!

In the mining industry, dynamite helped miners safely dig deeper mines faster. Sticks of dynamite were the perfect size and shape to be easily placed inside holes drilled in rock. Black powder was no longer needed. Today, black powder costs more than dynamite and is only used for special jobs.

Although Nobel's father had made explosives for use in war, Nobel hadn't created dynamite with war in mind. Still, dynamite ended up being used as a weapon in several wars. In the Spanish-American War, for example, it was used in cannons. It was also used by both sides fighting in World War I.

THE NOBEL PRIZES

NOBEL'S BROTHER LUDVIG DIED IN 1888, BUT NEWSPAPERS SAID NOBEL HIMSELF HAD DIED! ONE EVEN CALLED HIM "THE MERCHANT OF DEATH" BECAUSE OF HIS DANGEROUS WORK WITH EXPLOSIVES. THIS UPSET NOBEL. WHEN HE ACTUALLY DIED IN 1896, HE LEFT HIS MONEY FOR THE NOBEL PRIZES, WHICH HE'D SET UP TO HONOR PEOPLE WHO CHANGED THE WORLD IN GOOD AND PEACEFUL WAYS.

A quarry is an area where large amounts of rock, stone, sand, or other construction materials are taken out of the ground. Dynamite is sometimes used to blast out these materials.

INTERESTING AND STRANGE USES

Since its invention, dynamite has been put to many uses. It can be used to blast snow off mountainsides to prevent **avalanches**. Dynamite can also be detonated underwater, sending hundreds of dead fish to the surface to be collected by fishermen. This practice, called blast fishing, is illegal in many places but that doesn't stop people in some countries from doing it.

Dynamite has also been used to put out the flames if an oil well catches fire! This works because dynamite sucks up oxygen when it detonates, which removes the fire's fuel source so it can no longer burn.

NOT SO DYNAMITE

THERE ARE A FEW TYPES OF EXPLOSIVES THAT ARE COMMONLY CONFUSED WITH NOBEL'S DYNAMITE. PERHAPS THE ONE THAT'S MOST OFTEN MIXED UP IS TNT. TNT ISN'T AS POWERFUL AS DYNAMITE, BUT IT IS MORE STABLE. MILITARY DYNAMITE IS AN EXPLOSIVE MADE WITH TNT. AMMONIA DYNAMITE ISN'T MADE WITH NITROGLYCERIN.

DYNAMITE'S ODD JOBS

Dynamite is used for many interesting jobs. It can take down buildings without harming nearby structures. It can break up river ice so flooding doesn't occur when the ice begins to melt. It can also be used to blast tree stumps out of the ground.

AN UNCERTAIN FUTURE

However it may have happened, Alfred Nobel's accidental discovery over 150 years ago led to one of the greatest inventions of modern times. Nobel's dynamite proved to be a much safer—but still powerful—form of explosive.

Dynamite is still in use today, but the future of the explosive is unclear. In the mid-1800s, dynamite proved to be an incredible improvement within the explosives industry. In the time since its invention, it's been made even better. Dynamite is still used today for underwater blasting and certain demolition work. However, newer forms of advanced explosives have replaced it in many areas due to their increased stability.

Last One Standing

DYNO NOBEL, A COMPANY ESTABLISHED MORE THAN 100 YEARS AGO, IS THE ONLY MANUFACTURER OF DYNAMITE IN NORTH AMERICA, THOUGH IT PRODUCES MILLIONS OF POUNDS OF THE EXPLOSIVE EVERY YEAR. DYNAMITE REMAINS ONE OF THE MOST SUSTAINABLE FORMS OF EXPLOSIVE, WHICH MEANS ITS USE SHOULD BE ABLE TO CONTINUE FOR A LONG TIME.

Demolition experts work with explosives, like dynamite, to destroy things. For example, they might figure out how to use dynamite to bring down a whole row of buildings.

GLOSSARY

avalanche: a large mass of snow sliding down a mountain or over a cliff

chemist: a scientist who studies the structure and properties of substances and the changes that they go through

combustion: the act of burning

engineer: someone who uses science and math to build better objects

flammable: able to be set on fire and to burn quickly

industry: a group of businesses that provide a particular product or service

mechanical: relating to physical movement

molecule: a very small piece of matter

oxidizer: a type of chemical that a fuel needs in order to burn

patent: rights given to a person by the government to make, use, or sell an invention

silica: a chemical that contains silicon, that is found in sand and quartz, and that is used to make glass

subsonic: slower than the speed of sound

supersonic: faster than the speed of sound

volatile: likely to change in a very sudden or extreme way

wood pulp: a soft material that is made mostly from wood and is used in making paper

For More Information

BOOKS

Bearce, Stephanie. *Twisted True Tales from Science: Explosive Experiments*. Waco, TX: Prufrock Press Inc., 2017.

Devera, Czeena. *Alfred Nobel*. Ann Arbor, MI: Cherry Lake Publishing, 2019.

Verstraete, Larry. *Accidental Discoveries: From Laughing Gas to Dynamite*. Victoria, BC: FriesenPress, 2016.

WEBSITES

Alfred Nobel's Life and Work
www.nobelprize.org/alfred-nobel/alfred-nobels-life-and-work/
Learn more about Alfred Nobel, the scientist who discovered dynamite!

Dynamite
www.softschools.com/inventions/history/dynamite_history/370/
Read more fun facts about dynamite here.

How Does Dynamite Work?
wonderopolis.org/wonder/how-does-dynamite-work
Visit this site to find the answers to interesting questions about dynamite.

INDEX